The snake lies in the sun.
The sun keeps the snake warm.

Now the snake wriggles along.
It feels the ground
with its tongue.

WONDER
STARTERS

Snakes

Pictures by LORNA PAULL

Published by WONDER BOOKS
A Division of Grosset & Dunlap, Inc.

51 Madison Avenue New York, N.Y. 10010

Published in the United States by Wonder Books, a Division of Grosset & Dunlap, Inc.

ISBN: 0-448-09667-6 (Trade Edition)
ISBN: 0-448-06387-5 (Library Edition)

FIRST PRINTING 1973

Printed and bound in the United States.

Library of Congress Catalog Card Number: 73-1976

This is the snake's head.
It has no eyelids.
It has two big teeth in front.
It can fold the teeth back.

The snake finds a mouse.
It bites the mouse.
Poison from the big teeth
kills the mouse quickly.

The snake swallows the mouse.

The snake has a scaly skin.
The scales on its back are small.
The scales are bigger underneath.

Snakes often grow a new skin.
Then the old skin comes off.

In winter snakes find a hiding place.
They stay asleep.

Many snakes lay eggs.
Baby snakes soon come out.

This is a rattlesnake.
It lives in America.
It rattles as a warning sound.

This rattlesnake lives in a desert.
It wriggles sideways very quickly.

This is a python.
It lives in India.
It is very long.
It waits in a tree.

The python catches a goat.
It winds itself around the goat.
The goat cannot breathe.

This is a sea snake.
Sea snakes have flat tails.
Their tails help them swim.
They often eat fish.

This snake lives in trees.
It can glide to another tree.
It makes its body flat.
It is called a flying snake.

This snake eats eggs.
It swallows the inside of the egg.
It spits out the shell.

This snake is a cobra.
It lives in India.
It can make its neck very big.
It sometimes bites people.

This is a spitting cobra.
It can spit poison.

This man is a snake charmer.
He keeps his cobra in a basket.

Some people love snakes.
This man kept pet snakes.
They did not bite him.

Some animals hunt snakes.
The mongoose hunts snakes.
Some birds eat snakes.

See for yourself.

Go to the zoo.

How many snakes from this book can you find?

Starter's **Snakes** words

tongue
(page 2)

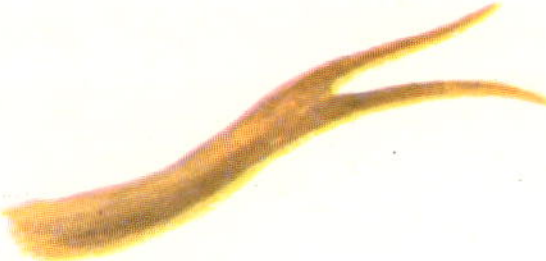

teeth
(page 3)

swallow
(page 5)

scales
(page 6)

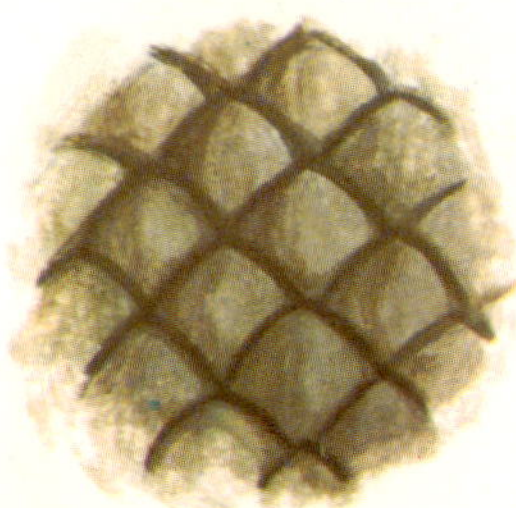

skin
(page 7)

egg
(page 9)

rattlesnake
(page 10)

America
(page 10)

desert
(page 11)

python
(page 12)

India
(page 12)

tree
(page 12)

goat
(page 13)

sea snake
(page 14)

tail
(page 14)

flying snake
(page 15)

glide
(page 15)

shell
(page 16)

cobra
(page 17)

neck
(page 17)

bite
(page 17)

snake charmer
(page 19)

basket
(page 19)

mongoose
(page 21)